Skill Builders Division

by Jessica Breur

Welcome to Rainbow Bridge Publishing's Skill Builders series. Like our Summer Bridge Activities collection, the Skill Builders series is designed to make learning both fun and rewarding.

Skill Builders Division provides students with focused practice to help them reinforce and develop division skills. Each Skill Builders volume is grade-level appropriate, with clear examples and instructions to guide the lessons. Skill Builders Division combines a number of approaches and activities to help students expand on their basic division skills with exercises on basic fact review, remainders, dividing up to 6-digit numbers by 3-digit divisors, estimation, dividing decimals and money, fractions, time conversions, averages, measurements, and more.

A critical thinking section includes exercises to help develop higher-order thinking skills.

Learning is more effective when approached with an element of fun and enthusiasm—just as most children approach life. That's why the Skill Builders combine entertaining and academically sound exercises with eye-catching graphics and fun themes—to make reviewing basic skills fun and effective, for both you and your budding scholars.

© 2005 Rainbow Bridge Publishing
All rights reserved
www.summerbridgeactivities.com

Table of Contents

Introduction 3

Basic Facts Review 4–8

Division Patterns 9–10

Division with Remainders 11–13

Reviewing Long Division 14–15

Dividing
   2-Digit Numbers 16–19
   3-Digit Numbers 20–21

Zeros in the Quotient 22

Dividing with
   Multiples of 10 23–24

To Estimate
   Is Really First Rate! 25

Dividing
   2 Digits by 2 Digits 26–28

Estimating with 3 Digits 29

Dividing
   3 Digits by 2 Digits 30–33

Dividing Money 34–35

Estimating with Money 36

Dividing
   4 Digits by 2 Digits 37–39
   5 Digits by 2 Digits 40–41

Using Number Sense 42

Dividing
   5 Digits by 3 Digits 43–44
   6 Digits by 3 Digits 45–46

What Is the Point? 47

Point Movers 48

Dividing Decimals
   by Decimals 49–50

On-the-Dot
   Decimal Practice 51–52

Fractions and Division 53–54

Dividing with
   Mixed Numbers 55

Dividing Whole Numbers
   by Fractions and
   Mixed Numbers 56

Dividing
   Fractions by Fractions 57

From Fraction to Percent 58

Algebra and Division 59–61

Time Conversions..................62–63

Measuring Up Your Division.......64

Metric Measures 65

What's Your Average? 66

Critical Thinking Skills
   Mission Possible 67
   Down on the Farm 68
   The Magic Quotient 69
   Breakfast Basics 70
   Pete's Precious Pies 71
   Division Riddles 72
   Divisibility Moves On 73
   The Divisibility Bowl 74

Answer Pages 75–80

Introduction

Division is a math operation using numbers to make a big number into a smaller number. There are three different ways to write a division problem, and each number in the problem has a special name.

A. The number you divide into is the **dividend**.

B. The number you divide by is the **divisor**.

C. The answer is also called a **quotient**.

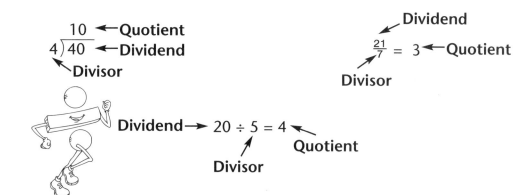

The following problems are solved for you. Practice labeling the numbers in a division problem as **quotient**, **divisor**, or **dividend**.

1. $30 \div 6 = \mathbf{5}$
   5 is the _quotient_

2. $18 \div 9 = 2$
   18 is the _____

3. $16 \div 8 = 2$
   16 is the _____

4. $42 \div 6 = 7$
   6 is the _____

5. $72 \div 8 = \mathbf{9}$
   9 is the _____

6. $28 \div 4 = 7$
   28 is the _____

Dividing Up to Three Is Easy— You'll See!

Knowing your basic facts is important for solving more difficult problems. Practice solving these problems.

Any number divided by 2 is half of that number.

2. $8 \div 2 =$ __4__
   $2 \div 2 =$ ____
   $24 \div 2 =$ ____
   $4 \div 2 =$ ____
   $6 \div 2 =$ ____
   $10 \div 2 =$ ____
   $14 \div 2 =$ ____
   $12 \div 2 =$ ____

Any number divided by 1 is the same number.

1. $2 \div 1 =$ __2__
   $8 \div 1 =$ ____
   $7 \div 1 =$ ____
   $1 \div 1 =$ ____
   $5 \div 1 =$ ____
   $3 \div 1 =$ ____
   $6 \div 1 =$ ____
   $10 \div 1 =$ ____

3. $6 \div 3 =$ ____
   $15 \div 3 =$ ____
   $12 \div 3 =$ ____
   $24 \div 3 =$ ____
   $3 \div 3 =$ ____
   $21 \div 3 =$ ____
   $9 \div 3 =$ ____
   $18 \div 3 =$ ____

www.summerbridgeactivities.com © Rainbow Bridge Publishing

Moving Up to Six Really Kicks!

4. 4 ÷ 4 = _____
12 ÷ 4 = _____
28 ÷ 4 = _____
36 ÷ 4 = _____
8 ÷ 4 = _____
44 ÷ 4 = _____
16 ÷ 4 = _____
20 ÷ 4 = _____

5. 5 ÷ 5 = _____
10 ÷ 5 = _____
30 ÷ 5 = _____
20 ÷ 5 = _____
15 ÷ 5 = _____
45 ÷ 5 = _____
60 ÷ 5 = _____
55 ÷ 5 = _____

6. 6 ÷ 6 = _____
24 ÷ 6 = _____
12 ÷ 6 = _____
54 ÷ 6 = _____
18 ÷ 6 = _____
72 ÷ 6 = _____
42 ÷ 6 = _____
30 ÷ 6 = _____

Solving Up to Nine... You're Right in Line!

7.
- 7 ÷ 7 = _____
- 21 ÷ 7 = _____
- 28 ÷ 7 = _____
- 63 ÷ 7 = _____
- 70 ÷ 7 = _____
- 49 ÷ 7 = _____
- 14 ÷ 7 = _____
- 84 ÷ 7 = _____

8.
- 8 ÷ 8 = _____
- 56 ÷ 8 = _____
- 32 ÷ 8 = _____
- 72 ÷ 8 = _____
- 48 ÷ 8 = _____
- 64 ÷ 8 = _____
- 96 ÷ 8 = _____
- 40 ÷ 8 = _____

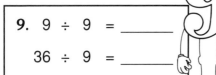

9.
- 9 ÷ 9 = _____
- 36 ÷ 9 = _____
- 27 ÷ 9 = _____
- 54 ÷ 9 = _____
- 18 ÷ 9 = _____
- 72 ÷ 9 = _____
- 45 ÷ 9 = _____
- 90 ÷ 9 = _____

Let's Delve into Twelve!

These can be done mentally.

Remember the zeros cancel each other out, so just look at the digits in the tens place.

11. 11 ÷ 11 = _____
55 ÷ 11 = _____
88 ÷ 11 = _____
99 ÷ 11 = _____
22 ÷ 11 = _____
66 ÷ 11 = _____
44 ÷ 11 = _____
33 ÷ 11 = _____

10. 10 ÷ 10 = _____
70 ÷ 10 = _____
50 ÷ 10 = _____
20 ÷ 10 = _____
60 ÷ 10 = _____
40 ÷ 10 = _____
90 ÷ 10 = _____
30 ÷ 10 = _____

12. 12 ÷ 12 = _____
48 ÷ 12 = _____
84 ÷ 12 = _____
60 ÷ 12 = _____
144 ÷ 12 = _____
96 ÷ 12 = _____
24 ÷ 12 = _____
36 ÷ 12 = _____

I give these problems a ten!

Word Problems Using the Basic Facts

Solve each problem.

1. Cameron has 16 horses lined up in groups of 4. How many groups do the horses make?

2. There are 21 cows in equal groups in 3 corrals. How many cows are in each corral?

3. Wesley was ready to shear the sheep. He made 5 groups out of 30 sheep. How many sheep are in each group?

4. Lori needed to gather the eggs from the chicken coop. She had 3 buckets to put 36 eggs in. How many eggs were in each bucket if each held the same number?

Division Pattern Review

Basic facts can help you divide big numbers.

Example:

    Basic Fact: 24 ÷ 6 = 4
                          240 ÷ 6 = 40
                       2,400 ÷ 6 = 400

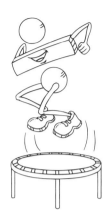

You simply solve the basic fact and add the zeros found in the dividend onto the quotient.

⚠ **CAUTION:**
    Be careful when there is a zero in the dividend that is part of the basic fact. Don't repeat that zero.

Example:

    Basic Fact: 40 ÷ 8 = 5
                          400 ÷ 8 = 50
                       4,000 ÷ 8 = 500

Try some on your own. Underline the basic fact to help you watch the zeros.

1.   14 ÷ 2 = 7
     140 ÷ 2 = ____
     1,400 ÷ 2 = ____

2.   32 ÷ 8 = ____
     320 ÷ 8 = ____
     3,200 ÷ 8 = ____

3.   20 ÷ 5 = ____
     200 ÷ 5 = ____
     2,000 ÷ 5 = ____

4.   90 ÷ 3 = ____
     900 ÷ 3 = ____
     9,000 ÷ 3 = ____

More Division Patterns

The basic facts and patterns will help you to divide mentally. Try this page mentally. Then check yourself with a calculator.

1.
   6 ÷ 3 = __2__
   60 ÷ 3 = __20__
   600 ÷ 3 = _____
   6,000 ÷ 3 = _____

2.
   36 ÷ 6 = _____
   360 ÷ 6 = _____
   3,600 ÷ 6 = _____
   36,000 ÷ 6 = _____

3.
   45 ÷ 5 = _____
   450 ÷ 5 = _____
   4,500 ÷ 5 = _____
   45,000 ÷ 5 = _____

4.
   12 ÷ 4 = _____
   120 ÷ 4 = _____
   1,200 ÷ 4 = _____
   12,000 ÷ 4 = _____

5. 50 ÷ 5 = _____

6. 210 ÷ 3 = _____

7. 1,800 ÷ 9 = _____

8. 80 ÷ 2 = _____

9. 200 ÷ 4 = _____

10. 420 ÷ 6 = _____

11. 8,100 ÷ 9 = _____

12. 2,500 ÷ 5 = _____

Basic Facts with Remainders

Sometimes when you try to divide a number of objects into groups of equal size, you have a few objects left over. The number of objects left over is called the **remainder**.

Example:

```
      7 r2
  8)58
    -56
      2
```

The quotient is 7 r2.

Try some for practice. Fill in the blanks to complete the algorithm.

1. 8)17
   –___

2. 3)8
   –___

3. 3)23
   –___

4. 2)5
   –___

5. 5)16
   –___

6. 9)32
   –___

7. 5)42
   –___

8. 6)14
   –___

9. 4)35
   –___

Basic Division with Remainders

Find each quotient and remainder.

1. $\overline{7\text{ r}1}$
    $3\overline{)22}$
    $\underline{-\ 21}$
    1

2. $6\overline{)39}$
   $\underline{}$

3. $5\overline{)36}$
   $\underline{}$

4. $4\overline{)33}$
   $\underline{}$

5. $5\overline{)27}$
   $\underline{}$

6. $8\overline{)50}$
   $\underline{}$

7. $7\overline{)59}$
   $\underline{}$

8. $9\overline{)37}$
   $\underline{}$

9. $2\overline{)3}$
   $\underline{}$

10. $8\overline{)66}$
    $\underline{}$

Word Problems with Remainders

Solve each problem.

1. Farmer Meghan has sixty-five rabbits. She builds nine cages. She wants to divide the rabbits up equally among the cages. How many rabbits will not fit in the cages?

2. Tanner buys seventy-six hay bales for the horses. He stacks them in piles of eight. How many hay bales are left over?

3. Kemry loves to watch the pigs. She sees that there are eight pigs, and only three at a time can fit at each of the two troughs. How many pigs can't eat at the same time as all the other pigs?

4. Allie wants to plant corn. She has made six rows, but she has forty seeds. How many seeds will she need to keep in order to plant each row with the same number of seeds?

Reviewing Long Division

Study Page

You have already been doing this with the basic facts of division, but let's take a closer look.

Example: $4\overline{)48}$

Step 1

$4\overline{)48}$

Decide if 4, the divisor, can go into the tens. Yes; 4 can make 1 group.

Step 2

$\begin{array}{r} 1 \\ 4\overline{)48} \end{array}$

Write the 1 above the 4 in the dividend.

Step 3

$\begin{array}{r} 1 \\ 4\overline{)48} \\ -4 \\ \hline 0 \end{array}$

Multiply 4 x 1, and subtract the product from the tens in the dividend.

Step 4

$\begin{array}{r} 1 \\ 4\overline{)48} \\ -4\downarrow \\ \hline 08 \end{array}$

Bring down the 8 ones.

Step 5

$\begin{array}{r} 1 \\ 4\overline{)48} \\ -4 \\ \hline 08 \end{array}$

See if 4, the divisor, can divide into the 8 from the dividend. Yes; it makes 2 groups.

Step 6

$\begin{array}{r} 12 \\ 4\overline{)48} \\ -4 \\ \hline 08 \end{array}$

Write the 2 above the 8 in the dividend.

Step 7

$\begin{array}{r} 12 \\ 4\overline{)48} \\ -4 \\ \hline 08 \\ -8 \\ \hline 0 \end{array}$

Multiply 2 x 4, and subtract the product from the ones in the dividend.

You are done because there are no more numbers in the dividend to work with.

Dividing into 2 Digits

Study Page

Let's try some different examples.

Example: 3)12

Step 1

3)12

Decide if 3, the divisor, can go into the 1. No.

Step 2

3)12

Decide if 3 can go into 12. Yes; 3 divides 12 into 4 groups.

Step 3

```
   4
3)12
```

Write the 4 above the 2 in the dividend.

Step 4

```
   4
3)12
 -12
   0
```

Multiply 4 x 3, and subtract the product from 12.

You are done since there are no more numbers to work with.

Example: 4)68

Step 1

4)68

Decide if 4 can go into the 6 in the dividend. Yes; it makes 1 group.

Step 2

```
   1
4)68
```

Write the 1 above the 6.

Step 3

```
   1
4)68
 - 4
   2
```

Multiply 4 x 1, and subtract the product from the 6.

Step 4

```
   1
4)68
 - 4↓
   28
```

Bring down the 8 ones in the dividend.

Step 5

```
   1
4)68
 - 4
   28
```

Since 2 is too small for 4 to divide, see how many times 4 goes into 28. It divides 28 into 7 groups.

Step 6

```
   17
4)68
 - 4
   28
 - 28
    0
```

Write the 7 above the 8. Multiply 7 x 4, and subtract the product from 28.

You are done.

Dividing 2-Digit Numbers

Long division uses multiple math operations, like multiplying, subtracting, and dividing. Work on writing the algorithm to show your steps.

1. 3)39

2. 3)84

3. 4)56

4. 5)60

5. 3)87

6. 8)96

7. 6)72

8. 7)98

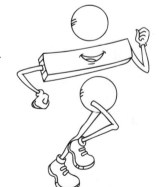

After working out the algorithm, check yourself with a calculator.

Dividing 2-Digit Numbers with a Remainder

Study Page

Example: $4\overline{)93}$

Steps 1 and 2

$$\begin{array}{r} 2 \\ 4\overline{)93} \end{array}$$

Decide if 4 can go into 9.
Yes; it can make 2 groups.
Write the 2 above the 9 in the dividend.

Steps 3 and 4

$$\begin{array}{r} 2 \\ 4\overline{)93} \\ -8\downarrow \\ \hline 13 \end{array}$$

Multiply 4 x 2, and subtract the product from the tens in the dividend.
Bring down the 3 to make a new dividend of 13.

Steps 5, 6, and 7

$$\begin{array}{r} 23 \\ 4\overline{)93} \\ -8 \\ \hline 13 \\ -12 \\ \hline 1 \end{array}$$

See how many times 4 goes into 13.
It goes into 13 three times.
Write 3 above the 3 in the dividend.
Multiply 4 x 3, and subtract the product from 13.

Step 8

$$\begin{array}{r} 23\ r1 \\ 4\overline{)93} \\ -8 \\ \hline 13 \\ -12 \\ \hline 1 \end{array}$$

Since there are no more numbers in the dividend to bring down, and the number left over (1) is smaller than the divisor, you have a remainder.

Dividing 2-Digit Numbers with a Remainder

Work the following problems.

1. 1☐ r☐
    7)85
    -7↓
      1☐
    -☐☐
       ☐

2. ☐☐ r☐
    6)76
    -6↓
      1☐
    -☐☐
       ☐

3. _)65

4. 6)69

5. 3)67

6. 8)99

7. 9)84

8. 4)89

9. 5)78

10. 2)27

Keep on Moving

Match the problems with their quotients.

1. 4)̄99   A. 8 r2
   = ____

2. 2)̄65   B. 22 r1
   = ____

3. 3)̄67   C. 9 r3
   = ____

4. 9)̄84   D. 24 r3
   = ____

5. 5)̄78   E. 32 r1
   = ____

6. 8)̄66   F. 15 r3
   = ____

Dividing 3-Digit Numbers

Sometimes you must solve a problem by looking at the first two digits, dividing the divisor into them, and then bringing down the third digit to complete the problem.

Example:

```
      64
   8)512
   -48
     32
    -32
      0
```

First divide 51 by 8 since 5 is too small to divide by 8.

Then bring down the 2 to make 32.

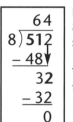

Try solving a few like the sample. Use a calculator to check your work.

1. 5)285

2. 3)984

3. 7)882

4. 6)210

5. 8)512

6. 9)972

Dividing 3-Digit Numbers with Remainders

Solve each problem.

1. 3)587

2. 9)948

3. 5)872

4. 6)424

5. 3)739

6. 9)838

7. 2)833

8. 4)211

Zeros in the Quotient

Sometimes there may not be enough tens or ones to divide into. If so, write 0 in the quotient over the proper number in the dividend. Then continue with the remaining steps to finish solving the problem.

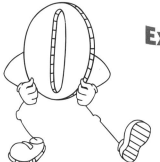

Example:

```
      105 r4
   6)634
    -6↓
      03
     - 0↓
       34
      -30
        4
```

3 is too small to divide by 6, so you put a 0 in the quotient, multiply, and subtract.

Fill in the blanks to solve these problems

1. 3)901

2. 4)435

3. 2)619

4. 8)324

To Estimate Is Really First Rate!

Estimating is a skill that people use daily. You estimate how much toothpaste to put on your toothbrush or how much time you need to get ready before school. In division, estimation can make bigger problems easier.

Now that you have tried a few estimation problems with multiples of 10, you can apply that skill to 2-digit divisors that are close to a multiple of 10. We will call this type of estimation front-end estimation because we will focus on the first digit in both the divisor and the dividend.

Example: 12)53

Step 1   First, estimate how many 12s get you close to 53, or think how many 10s are close to 50 (10 x 5). This is called front-end estimation. You use the first digit in the numbers to help you get close to the number you will really need to solve the equation.

Step 2   Since 12 is larger than 10*, you should move down a digit from 5 to 4 to help you solve the equation.

Step 3   Place the 4 above the correct number in the dividend, and solve.

$$\begin{array}{r} 12 \\ \times\ 4 \\ \hline 48 \end{array}$$

$$\begin{array}{r} 4\ r5 \\ 12\overline{)53} \\ -48 \\ \hline 5 \end{array}$$

Let's see if you can fill in the blank using a number to help you estimate.

1. 88 ÷ 22   Think: 80 ÷ _____
2. 54 ÷ 18   Think: 50 ÷ _____
3. 87 ÷ 29   Think: _____ ÷ 30
4. 95 ÷ 19   Think: 90 ÷ _____
5. 78 ÷ 26   Think: _____ ÷ 20

*Sometimes when you multiply, your estimate will be too high. When that happens, try a lower estimate.

Dividing a 2-Digit Number by a 2-Digit Divisor

You can use front-end estimation or your basic facts to help find the correct number to put in the quotient.

1. 25)75   (Hint: 70 ÷ 20)

2. 17)68   (Hint: 60 ÷ 10)

3. 23)92

4. 13)91

5. 15)75

6. 14)84

7. 26)78

8. 13)65

9. 15)90

10. 11)77

11. 19)95

12. 16)80

Dividing 2-Digit Dividends with a Remainder

Try estimating to help you find the best number for solving the equation.

1. 20)74

2. 24)98

3. 35)75

4. 22)99

5. 15)81

6. 33)85

7. 16)66

8. 11)73

9. 22)99

10. 44)92

11. 23)51

12. 29)89

Match It Up: More Practice Dividing by 2-Digit Divisors

Estimate to mentally figure out the answer, and then match it to an answer on the right. Check your match with a calculator.

1. 46)564   A. 3 r18

2. 52)167   B. 24

3. 26)96   C. 4 r5

4. 22)226   D. 12 r12

5. 12)53   E. 3 r33

6. 14)94   F. 7 r7

7. 42)159   G. 5 r15

8. 16)384   H. 6 r10

9. 25)140   I. 3 r11

10. 17)126   J. 10 r6

Estimating when Dividing 3 Digits by 2 Digits

To estimate with a 3-digit dividend, you use the tens place in the dividend to help you estimate. In fact, you may have to use the first 3 digits.

Example: 23)‾161

Step 1   Think 20)‾160   (16 ÷ 2) Eight 20s will go into 160, but 8 × 23 = 184, which is too high. Lower your estimate to 7.

Step 2   Place the 7 above the 1.
Multiply, and subtract from the dividend.

$$\begin{array}{r}7\\23\overline{)161}\\-161\\\hline 0\end{array}$$

Let's see if you can fill in the blanks using a number to help you estimate.

1. 267 ÷ 21   Think: 260 ÷ _____

2. 161 ÷ 23   Think: 160 ÷ _____

3. 231 ÷ 42   Think: _____ ÷ 40

4. 852 ÷ 93   Think: 850 ÷ _____

5. 353 ÷ 55   Think: _____ ÷ 50

Dividing 3 Digits by 2 Digits

Try solving these problems.
Knowing your basic facts will help you go faster.

1. 40)880   (Hint: 80 ÷ 40)   7. 52)988

2. 33)957   (Hint: 90 ÷ 30)   8. 21)966

3. 23)828   9. 37)925

4. 18)936   10. 30)810

5. 15)960   11. 60)720

6. 44)880   12. 54)972

Dividing 3 Digits by 2 Digits with Remainders

Try solving these problems.
Knowing your basic facts will help you go faster.

1. 22)687

2. 23)594

3. 47)947

4. 50)999

5. 43)820

6. 34)920

7. 62)832

8. 15)874

9. 58)936

10. 92)999

11. 90)986

12. 18)674

Division Puzzle

Work these problems, and use the answers to help you fill in the blanks.

Across

1. 728 ÷ 14 = _____
3. 527 ÷ 17 = _____
5. 869 ÷ 11 = _____
6. 931 ÷ 19 = _____
7. 945 ÷ 15 = _____
9. 728 ÷ 26 = _____
11. 903 ÷ 43 = _____
12. 884 ÷ 26 = _____

Down

1. 684 ÷ 12 = _____
2. 464 ÷ 16 = _____
3. 850 ÷ 25 = _____
4. 969 ÷ 51 = _____
7. 744 ÷ 12 = _____
8. 961 ÷ 31 = _____
9. 782 ÷ 34 = _____
10. 924 ÷ 11 = _____

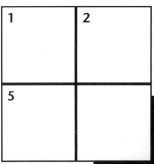

Word Problems with 3-Digit Dividends

Hint: Remember to include the remainder.

1. Farmer Kemry prints 250 fliers to advertise that her produce is on sale after the harvest. About how many boxes will she need if she puts 80 fliers in a box? How many fliers will be in the last box?

2. A chicken farmer uses egg cartons made from recycled materials. If 6 eggs fit into each carton, how many cartons will he need for 350 eggs? How many eggs will be in the last carton?

3. Farmer Denise makes cheese from her famous cows. She makes 950 pounds of cheese. How many crates will she need if each crate holds 50 pounds?

4. The sheep have just been sheared. Farmer Wesley wants to sell 893 pounds of wool to 19 stores. How many pounds will each store get to make sweaters from his wool? Will there be any wool left over? How much?

Dividing Money

Dividing money is the same as dividing whole numbers. We add a decimal point to our dividends, but the numerals in the quotients do not change.

Example:

```
         36              $0.36
    18)648           18)$6.48
      -54              -54
       108              108
      -108             -108
         0                0
```

Try some on your own. Remember to place the decimal point in the quotient directly above the decimal point in the dividend.

1. 25)$12.50

2. 18)$7.92

3. 75)$32.25

4. 15)$24.00

5. 34)$31.28

6. 53)$66.25

More Dividing Money

Remember to include the decimal point and money symbol as you solve these problems.

1. 23)$32.43

2. 13)$8.97

3. 35)$537.60

4. 80)$74.40

5. 11)$12.54

6. 21)$8.40

7. 12)$5.64

8. 17)$6.80

9. 34)$36.38

10. 16)$8.00

11. 30)$66.90

12. 37)$99.16

Estimating with Money

Use your number sense ability to help you choose the best answer. Circle the best answer.

1. $40.00 ÷ 23 is   more than $1.00   less than $1.00

2. $9.09 ÷ 9 is   more than $1.00   less than $1.00

3. $10.98 ÷ 12 is   more than $1.00   less than $1.00

4. $21.00 ÷ 23 is   more than $1.00   less than $1.00

5. $74.40 ÷ 80 is   more than $1.00   less than $1.00

6. $32.43 ÷ 23 is   more than $1.00   less than $1.00

7. $9.00 ÷ 16 is   more than $1.00   less than $1.00

8. $127.00 ÷ 50 is   more than $1.00   less than $1.00

9. $66.90 ÷ 29 is   more than $1.00   less than $1.00

10. $8.40 ÷ 22 is   more than $1.00   less than $1.00

Dividing 4 Digits by 2 Digits

We are now adding more numbers to your dividend. Just keep following the same steps until your dividend is smaller than the divisor. Some problems will have a remainder; others will not.

Example:

(Hint: 100 ÷ 50)

```
        203 r13
   49)9,960
       -9 8
         16
        - 0
        160
       -147
         13
```

Try some on your own.

1. 29)8,971   (Hint: 80 ÷ 20 = 4, so move down to 3)

2. 40)8,413

3. 48)9,851

4. 22)9,020

5. 52)1,700

6. 19)9,158

More Dividing 4 Digits by 2 Digits

1. 50)2,250

2. 41)1,450

3. 29)2,436

4. 78)6,890

5. 31)2,356

6. 62)2,984

7. 24)7,920

8. 90)6,480

9. 35)7,000

10. 37)8,658

Even More Practice Dividing 4 Digits by 2 Digits

1. 73)5,402

2. 34)7,301

3. 44)8,579

4. 28)7,548

5. 56)9,968

6. 39)6,000

7. 18)7,128

8. 20)8,740

9. 88)7,850

10. 67)8,752

Dividing 5 Digits by 2 Digits

The bigger the dividend, the more times you must repeat the steps to complete the problem. Try a few after studying the example.

Example:

```
        450 r64
    74)33,364
      -296
        376
       -370
         64
```

You are becoming a whiz at these!

1. 49)10,058

2. 66)26,850

3. 75)14,230

4. 34)90,930

More Practice with Dividends with 5 Digits

1. 90)68,892

2. 53)18,496

3. 19)46,854

4. 88)60,900

5. 37)20,424

6. 85)92,727

7. 26)85,000

8. 44)24,791

9. 81)77,077

10. 72)89,496

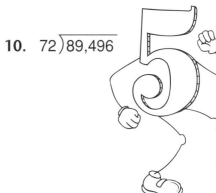

Using Number Sense with Division

In the following examples, a quotient is given. Then there is an explanation of whether that quotient is possible or not.

Example:

$$32\overline{)22{,}528}^{\,74}$$

$$25\overline{)76{,}000}^{\,3{,}040}$$

Not possible because 225 can be divided by 32.

Possible since 76 is divisible by 25 (3 times).

Look at the following problems, and decide if the quotient given is possible or not based on your ability to look at the divisor and the dividend.

1. $58\overline{)32{,}456}^{\,55}$ possible   not possible

2. $46\overline{)38{,}429}^{\,835}$ possible   not possible

3. $76\overline{)53{,}653}^{\,1{,}705\ r73}$ possible   not possible

4. $30\overline{)94{,}786}^{\,3{,}159\ r16}$ possible   not possible

Dividing 5 Digits by a 3-Digit Divisor

When dividing by a 3-digit divisor, you can still use front-end estimation on the first 3 or 4 digits in the dividend.

Example:

$$\begin{array}{r} 3 \\ 250 \overline{)95{,}500} \\ -750 \\ \hline 205 \end{array}$$

Finish solving the rest of the problem.

$$\begin{array}{r} 382 \\ 250 \overline{)95{,}500} \\ -750 \\ \hline 2050 \\ -2000 \\ \hline 500 \\ -500 \\ \hline 0 \end{array}$$

900 can be divided by 200 about 4 times.
But 250 is greater than 200, so use 3.

Solve a few on your own.

1. $524 \overline{)87{,}508}$

2. $362 \overline{)91{,}000}$

3. $674 \overline{)58{,}792}$

4. $434 \overline{)99{,}386}$

More Practice Dividing 5 Digits by 3 Digits

1. 680) 40,187

2. 309) 88,374

3. 268) 94,977

4. 574) 38,500

5. 737) 43,746

6. 902) 66,008

7. 233) 94,973

8. 837) 79,439

Dividing 6 Digits by 3 Digits

You will follow the same first step as you do if dividing into 5 digits. You will need to look at the first 3 to 4 digits in the dividend to see where to place the first number in the quotient.

Example:

$$583 \overline{)413{,}499}^{7}$$

Solve the rest of the problem.

$$583 \overline{)413{,}499}^{709\ r152}$$
$$\underline{-4081}$$
$$539$$
$$\underline{-0}$$
$$5399$$
$$\underline{-5247}$$
$$152$$

500 can divide 4,000 8 times.
But 583 is greater than 500, so use 7.

Try a few on your own.

1. $446 \overline{)383{,}560}$

2. $509 \overline{)309{,}981}$

3. $923 \overline{)800{,}000}$

4. $692 \overline{)642{,}946}$

Dividing 6 Digits by 3 Digits

Solve.

1. 540) 756,000

4. 811) 702,010

2. 458) 415,864

5. 738) 477,500

3. 372) 203,112

6. 251) 176,955

What Is the Point?

Dividing a decimal by a whole number is just like dividing money. But sometimes you need to add a zero to your dividend to help you finish the problem without a remainder.*

Example:

$$3 \overline{) 50.4} = 16.8$$

```
    16.8
3) 50.4
   -3
   20
  -18
   24
  -24
    0
```

Remember to put the decimal up in the quotient.
Then divide.

```
     5.82
5) 29.10
  -25
   41
  -40    Add a zero.
   10
  -10
    0
```

Try a few on your own.

1. $1.6 \overline{) 25.92}$

2. $9 \overline{) 5.166}$

3. $21 \overline{) 92.001}$

4. $7 \overline{) 41.867}$

5. $57 \overline{) 202.35}$

6. $72 \overline{) 263.70}$

*Using a decimal is actually a way of writing a remainder in a different form.

Point Movers

Division patterns are easy when dividing by 10, 100, and 1,000—you just add on zeros. But when dividing decimals by multiples of 10, you can just move the decimal point to obtain your quotient. Move the decimal to the left for as many zeros as there are in the divisor. You can do it mentally. (Hint: You may have to add zeros to help you with place value.)

Example: 19.38

19.38 ÷ 1<u>0</u> = 1.938   19.38 ÷ 1<u>00</u> = .1938   19.38 ÷ 1,<u>000</u> = .01938

Move the decimal <u>once</u> to the left.

Move the decimal <u>twice</u> to the left.

Move the decimal <u>three</u> times to the left; add a 0 if needed.

	÷10	÷100	÷1,000
1. 453.2			
2. 7.102			
3. 986.03			
4. 206			
5. 545.9			
6. 1,605.83			
7. 36.25			
8. 682.4			
9. 2.068			
10. 13			

Dividing Decimals by Decimals

Dividing by a decimal is like dividing by a whole number. However, when you divide by a decimal, you must move the decimal over in the divisor to make a whole number and then move the decimal point in the dividend the same number of spaces. You may have to add zeros to the dividend help with place value.

Example:

$$1.6\overline{).768}$$

Step 1 Change the divisor to a whole number by moving the decimal point to the right.

Step 2 Move the decimal in the dividend the same number of spaces to the right, and put a decimal up in the quotient above the decimal in the dividend.

$$16\overline{)7.68}$$

Step 3 Then divide.

$$\begin{array}{r} .48 \\ 16\overline{)7.68} \\ -64 \\ \hline 128 \\ -128 \\ \hline 0 \end{array}$$

Try a few on your own.

1. $2.2\overline{)8.36}$

2. $.03\overline{).48}$

3. $.8\overline{)6.016}$

4. $.16\overline{)72}$

5. $5.4\overline{)39.42}$

6. $32\overline{)185.6}$

You're a Decimal Divider

1. $1.2 \overline{)4.2}$

2. $0.5 \overline{)8.5}$

3. $0.2 \overline{)0.0006}$

4. $1.2 \overline{)24.6}$

5. $6.8 \overline{)14.552}$

6. $0.2 \overline{)0.5}$

7. $2.03 \overline{)3.248}$

8. $0.09 \overline{)8.1}$

9. $1.25 \overline{)2.75}$

10. $6.5 \overline{)0.13}$

On-the-Dot Decimal Practice

Insert a decimal point to make the equation true.

1. 7.4096 ÷ 1.1 = 6 7 3 6
2. 22.3443 ÷ 5.49 = 4 0 7
3. 3.1515 ÷ 1.5 = 2 1 0 1
4. 4.8225 ÷ 2.5 = 1 9 2 9
5. 17.22 ÷ 4.1 = 4 2 0

6. 12.576 ÷ 2.4 = 5 2 4
7. 6.684 ÷ 0.06 = 1 1 1 4
8. 3.48 ÷ 5.8 = 0 6
9. 87.4 ÷ 0.38 = 2 3 0
10. 6.89 ÷ 1.3 = 5 3

Match each problem to the quotient that makes it true.

11. 2.74) 0.685
12. 6.1) 0.427
13. 8.2) 126.28
14. 0.71) 0.8449
15. 60.3) 422.1

A. 7
B. 0.07
C. 15.4
D. 1.19
E. 0.25

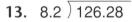

Word Problems Using Decimals

Solve each problem.

1. Mrs. Breur's class is going to a matinee movie about farm animals. The class of 36 students will share the cost of the movie. How much will each one pay if the total cost for the movie is $95.40?

2. A square fence around the corn has a perimeter of 10.248 m. How long is each side?

3. Tanner bought 3 lbs of wheat to feed the chickens. It cost $75.42. How much would just 1 lb cost?

4. Farmer Cameron wants to turn garbage into dirt! He puts table scraps in his worm bin to create soil that is great for his garden. He wants to divide 9.2 pounds of garbage equally among the 4 sections of the worm bin. How much garbage should be in each bin?

Fractions and Division

Dividing by a fraction is the same as multiplying by its reciprocal. A **reciprocal** is a fraction whose numerator and denominator have been switched.

Example: (Hint: The reciprocal of $\frac{2}{3}$ is $\frac{3}{2}$.)

If Denise wants to make a recipe that makes about 6 cups of soup, and she wants to know how many $\frac{2}{3}$-cup servings this recipe will make, she will need to think of dividing the 6 cups of soup into equal groups of $\frac{2}{3}$, or $6 \div \frac{2}{3}$.

To find the answer, you must first multiply the 6 by the denominator (3) and then divide by the numerator (2).

$6 \div \frac{2}{3} =$ The reciprocal of $\frac{2}{3}$ is $\frac{3}{2}$.

$\frac{6}{1} \times \frac{3}{2} =$ The whole number 6 is written as the fraction $\frac{6}{1}$.

$\frac{6}{1} \times \frac{3}{2} = \frac{18}{2}$

$\frac{18}{2} = 9$ Simplify the fraction by dividing 18 by 2 to get 9.

Try a few on your own. Some of the steps are done for you.

1. $4 \div \frac{1}{2} = \frac{4}{1} \times \frac{2}{1} = $ _____

2. $3 \div \frac{3}{5} = \frac{3}{\_\_} \times \frac{5}{\_\_} = \frac{15}{\_\_} = 5$

3. $5 \div \frac{1}{4} = \frac{5}{1} \times $ _____ = _____

4. $8 \div \frac{1}{7} = $ _____ × _____ = _____

5. $4 \div \frac{3}{4} = $ _____ × _____ = _____ = _____

6. $12 \div \frac{2}{5} = $ _____ × _____ = _____ = _____

More Dividing with Fractions

Sometimes your answer must be given as a mixed number or as a fraction instead of a whole number.

Example: $4 \div \frac{3}{4} = \frac{4}{1} \times \frac{4}{3} =$ Get the reciprocal; then multiply.

$\frac{16}{3}$ Simplify by dividing 16 by 3 and writing the remainder above the denominator.

$5\frac{1}{3}$

Try a few on your own.

1. $1 \div \frac{4}{7} =$ _____

   $\frac{1}{1} \times \frac{7}{4} = \frac{7}{4} = 1$ ___/4

2. $11 \div \frac{3}{4} =$ _____

   $\frac{11}{1} \times \frac{4}{3} =$ _____ $= 14 \,^2\!/$___

3. $3 \div \frac{2}{5} =$ _____

4. $8 \div \frac{1}{3} =$ _____

5. $6 \div \frac{1}{2} =$ _____

6. $8 \div \frac{3}{7} =$ _____

7. $7 \div \frac{2}{9} =$ _____

8. $7 \div \frac{3}{10} =$ _____

9. $5 \div \frac{2}{3} =$ _____

10. $10 \div \frac{17}{4} =$ _____

Dividing with Mixed Numbers

To divide by a mixed number, rewrite the mixed number as an improper fraction. Then multiply by the reciprocal.

Example: $5 \div 2\frac{1}{4} =$

<u>Step 1</u>　　　　$2\frac{1}{4} = \frac{9}{4}$

To make the improper fraction, multiply the denominator by the whole number, and then add the numerator: (2 x 4) + 1 = 9. Place your answer over the original denominator (4).

<u>Step 2</u>

$\frac{5}{1} \div \frac{9}{4} =$　　　Multiply by the reciprocal.　　$\frac{5}{1} \times \frac{4}{9} = \frac{20}{9}$

<u>Step 3</u>

$\frac{20}{9} = 2\frac{2}{9}$　　　Simplify by dividing 9 into 20 and placing the remainder (2) over the denominator (9).

Try some on your own.

1. $5 \div 2\frac{2}{5} = \frac{5}{1} \times {}^5\!/\!\underline{} =$

　　$\underline{}/12 = \underline{}$

2. $8 \div 2\frac{1}{4} = \underline{}$

3. $15 \div 1\frac{2}{3} = \underline{}$

4. $2 \div 3\frac{1}{5} = \underline{}$

5. $12 \div 2\frac{2}{5} = \underline{}$

6. $5 \div 2\frac{2}{9} = \underline{}$

7. $8 \div 1\frac{1}{4} = \underline{}$

8. $3 \div 2\frac{3}{5} = \underline{}$

9. $7 \div 2\frac{4}{7} = \underline{}$

10. $10 \div 1\frac{7}{9} = \underline{}$

Dividing Whole Numbers by Fractions and Mixed Numbers

1. $6 \div \frac{1}{3} =$ _____

2. $10 \div 7\frac{2}{3} =$ _____

3. $7 \div \frac{6}{5} =$ _____

4. $8 \div 8\frac{7}{8} =$ _____

5. $7 \div 6\frac{3}{4} =$ _____

6. $3 \div \frac{6}{7} =$ _____

7. $4 \div 3\frac{5}{8} =$ _____

8. $8 \div 2\frac{1}{6} =$ _____

9. $1 \div \frac{3}{4} =$ _____

10. $5 \div \frac{9}{2} =$ _____

11. $2 \div 4\frac{2}{7} =$ _____

12. $16 \div \frac{2}{5} =$ _____

The digits are really coming down today!

Dividing Fractions by Fractions

When you divide a fraction by a fraction, you get the same answer as if you had multiplied the first fraction by the reciprocal of the second fraction.

Example: $\frac{2}{3} \div \frac{1}{12} =$

$\frac{2}{3} \times \frac{12}{1} = \frac{24}{3} = 8$

Try a few on your own.

1. $\frac{3}{6} \div \frac{1}{12} = $ _____

2. $\frac{1}{2} \div \frac{1}{4} = $ _____

3. $\frac{2}{3} \div \frac{1}{6} = $ _____

4. $\frac{2}{4} \div \frac{2}{12} = $ _____

5. $\frac{4}{5} \div \frac{5}{8} = $ _____

Remember to convert your mixed number to an improper fraction first to solve the following equations.

6. $\frac{2}{5} \div 3\frac{1}{4} = $ _____

7. $\frac{2}{3} \div 2\frac{2}{3} = $ _____

8. $\frac{1}{2} \div 1\frac{1}{3} = $ _____

9. $\frac{5}{8} \div 1\frac{1}{4} = $ _____

10. $\frac{1}{4} \div 1\frac{1}{4} = $ _____

Who's ready for 1/6 of a pizza?

Converting Fractions into Percents

We use division to help find percentages all the time. Follow the example to help you solve the problems.

Convert the following fraction to a percentage.

Example: $\frac{1}{3}$

<u>Step 1</u>
Divide the denominator into the numerator. Remember to add a decimal and zeros onto your whole number to complete the problem.

```
      .33
  3)1.00
    - 9
     10
    - 9
      1
```

<u>Step 2</u>
Round to the nearest hundredth.   .33

<u>Step 3</u>
Move the decimal 2 places to the right, and add the percent sign (%).   **33%**

Try some on your own.

1. $\frac{1}{5}$ = _____   6. $\frac{2}{5}$ = _____

2. $\frac{1}{4}$ = _____   7. $\frac{1}{10}$ = _____

3. $\frac{3}{4}$ = _____   8. $\frac{9}{20}$ = _____

4. $\frac{4}{7}$ = _____   9. $\frac{5}{6}$ = _____

5. $\frac{2}{3}$ = _____   10. $\frac{1}{2}$ = _____

Division in Algebra

In algebra, you solve an equation for an unknown quantity. You must "balance" your equation, and division will help. Study the following example, and then try some on your own.

Example: $8m = 72$

$$\frac{8m}{8} = \frac{72}{8}$$

Divide both sides of the equation by 8 to balance it.
$8 \div 8 = 1$, so m stands by itself.
$m = 72 \div 8 = 9$

$m = 9$

Your turn. Mental math and your knowledge of decimals will help you.

1. $6a = 24$

2. $3m = 45$

3. $\$0.15b = \$0.75$

4. $3j = 2.1$

5. $0.4w = 2.4$

6. $0.3y = 0.9$

$a + b = c$

Division and Algebra

Match the equation to the answer.

1. $4x = 68$ A. 6

2. $9p = 99$ B. 23

3. $21n = 168$ C. 4

4. $43t = 172$ D. 8

5. $18j = 216$ E. 26

6. $9x = 234$ F. 11

7. $5d = 115$ G. 44

8. $2c = 88$ H. 12

9. $54x = 162$ I. 17

10. $6x = 36$ J. 3

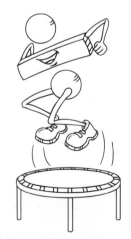

Division and Algebra

Try a few more algebra problems.

1. $1.2y = 3.6$

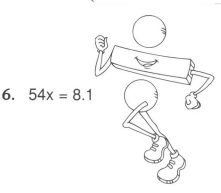

6. $54x = 8.1$

2. $5q = 4.0$

7. $8.73t = 31.428$

3. $4.2t = 4.62$

8. $6k = 34.8$

4. $0.18n = 5.4$

9. $0.5x = 3.5$

5. $0.11n = 0.275$

10. $0.6p = \$9.60$

Time Conversions

Division will help you figure out increments of time if you know the following information.

60 seconds = 1 minute	7 days = 1 week
60 minutes = 1 hour	4 weeks = 1 month
24 hours = 1 day	12 months or 52 weeks = 1 year

Example: 180 minutes = 3 hours (180 ÷ 60 = 3)

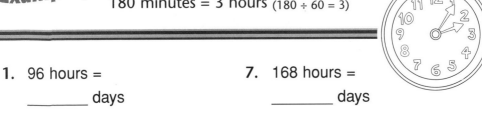

1. 96 hours = _____ days

2. 120 minutes = _____ hours

3. 300 seconds = _____ minutes

4. 49 days = _____ weeks

5. 48 hours = _____ days

6. 240 minutes = _____ hours

7. 168 hours = _____ days

8. ⅙ day = _____ hours

9. ⅓ minute = _____ seconds

10. 16 weeks = _____ months

11. 104 weeks = _____ years

12. 48 months = _____ years

Converting Time with a Remainder

Even when figuring out different increments of time, you may have leftover parts. Use the chart to help you solve the problems.

60 seconds = 1 minute	7 days = 1 week
60 minutes = 1 hour	4 weeks = 1 month
24 hours = 1 day	12 months or 52 weeks = 1 year

Example: 72 seconds = __1__ minute __12__ seconds

(72 ÷ 60 = 1 r12)

1. 132 seconds =

    ____ minutes

    ____ seconds

4. 78 hours =

    ____ days

    ____ hours

7. 50 days =

    ____ weeks

    ____ day

2. 26 hours =

    ____ day

    ____ hours

5. 16 months =

    ____ year

    ____ months

8. 314 seconds =

    ____ minutes

    ____ seconds

3. 189 minutes =

    ____ hours

    ____ minutes

6. 39 weeks =

    ____ months

    ____ weeks

9. 54 hours =

    ____ days

    ____ hours

Measuring Up Your Division Skills

Use the charts to help you solve these measurement problems.

12 inches = 1 foot	3 feet = 1 yard
5,280 feet = 1 mile	1,760 yards = 1 mile

1. 9 feet = _____ yards
2. 5,280 yards = _____ miles
3. 15,840 feet = _____ miles
4. 8,888 yards = _____ miles
5. 216 inches = _____ feet
6. 10,560 feet = _____ miles

1 pound (lb) = 16 ounces (oz)	1 ton (T) = 2,000 lb

1. 8,000 lb = _____ T
2. 240 oz = _____ lb
3. 1,920 oz = _____ lb
4. 60,000 lb = _____ T

1 pint = 2 cups	1 quart = 2 pints	1 gallon = 4 quarts

1. 8 quarts = _____ gallons
2. 16 pints = _____ quarts
3. 24 cups = _____ pints
4. 100 quarts = _____ gallons

Metric Measures

Use the charts to help you convert the metric units of measure.

| 1 cm = 10 mm | 1 m = 100 cm | 1 km = 1,000 m |

1. The farm is 8,000 m long.
   How many kilometers is that?
   8,000 m = _____ km

2. The robin's egg is 40 mm long.
   How long is that in centimeters?
   40 mm = _____ cm

3. The horseshoe was 190 mm wide.
   How many centimeters is that?
   190 mm = _____ cm

4. The bull in the corral was 300 centimeters long.
   That is how many meters?
   300 cm = _____ m

5. The sheep's tail was 70 millimeters long.
   How many centimeters is that?
   70 mm = _____ cm

What's Your Average?

You find an **average** by dividing the sum of a group of numbers by the number of addends. For example, to find the average of 21, 34, and 44, you would add the numbers up and divide the total by 3 since there are three addends.

21 + 34 + 44 = 99 ÷ 3 = 33, so 33 is the average.

Find the averages, and match each problem with its answer.

1. 85 + 200 + 100 + 95 =   A. 262

2. 81 + 57 + 63 =   B. 67

3. 96 + 120 + 102 + 124 + 95 + 125 =   C. 27

4. 105 + 100 + 104 =   D. 120

5. 278 + 246 =   E. 103

6. 35 + 26 + 18 + 25 + 31 =   F. 110.33

Mission Possible: Find the Missing Digits

Critical Thinking Skills

In the problems below, four digits are shown in the box. Fill in each blank with one of the digits to arrive at the correct quotient shown. Use estimation and guess/check to help you choose combinations of the digits that might work. Use a calculator to check your guesses.

1. 6, 4, 8, 9   __) __ __ __ = 9 4

2. 3, 5, 8, 7   __) __ __ __ = 5 5

3. 6, 2, 2, 8   __) __ __ __ = 4 7

4. 2, 8, 8, 9   __) __ __ __ = 9 2

5. 7, 4, 4, 1   __) __ __ __ = 6 3

6. 0, 2, 3, 7   __) __ __ __ = 2 9

Critical Thinking Skills

Down on the Farm

Your science class visits Mr. Breur's farm to learn about the animals there. Use basic facts and division to help you answer these questions.
(You may need to draw pictures to help you.)

1. There are 42 pigs on the farm, and each pigpen holds either 7 or 14 pigs. What is the greatest number of pigpens Mr. Breur needs?

2. Mr. Breur collects 32 eggs from the henhouse once a week. If each hen lays 2 or 4 eggs a week, what is the most hens Mr. Breur could have?

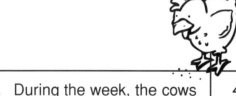

3. During the week, the cows roam free on the farm. But Mr. Breur puts the cows in the corral on the weekend. He has 80 cows on his farm, and a corral can hold 16 to 20 cows. What is the least number of corrals Mr. Breur needs?

4. Mr. Breur's neighbor has only ducks and pigs on his farm. There are 22 animals in all. Together, the animals have 58 legs. How many ducks does the farmer have?

The Magic Quotient

You can get your answer to match the number you started with by following these steps:

1. Enter any number with 3 digits into the calculator.
2. Repeat the 3 digits to make a 6-digit number.
3. Divide the number by 13.
4. Divide the result by 11.
5. Now divide by 7, and your answer should be the same number you started with.

Use your calculator to try these out.

Try your own.

3-digit number	239	172	363	_____	_____
Numbers with digits repeated	239,239	172,172			
÷ 13	18,403				
÷ 11	1,653				
÷ 7	239				

Can you figure out why this works?

Critical Thinking Skills

Breakfast Basics

Rachel needs to buy cereal for her family of six. Each person eats about four ounces of cereal a day. She needs to buy enough for two days. Decide which cereal to buy.

$3.00

$3.95

$3.45

1. First, find the cost for each ounce of cereal. You may have to round to the nearest cent.
   Munchie Crunch _____
   Frosty Bran Flakes _____
   Cracklin' Corn Pops _____

2. Which cereal is the best buy? _____

3. What is the least amount of money Rachel can spend on cereal and still have enough cereal for her family to eat? How many boxes is that? _____

Pete's Precious Pies

Pete believes that pies have become dull and boring. He has created new pies to put some flavor back into your taste buds.

1. Pete's specialty is Flaming Pies. He added a jar of hot peppers to the recipe. The pies were so hot customers could only eat $\frac{2}{9}$ of the pie. How many customers can he serve with 16 of these pies?

2. Pete gave away $\frac{2}{7}$ of a Crawling Creatures Pie to different customers until all 12 of these pies were gone. How many customers received a piece of free pie?

3. Each member of your soccer team was treated to $\frac{3}{7}$ of a Marvelous Mud Pie after a rainy game. It took 6 pies to treat all of you. How many members were on your team?

4. Each slice of Cantaloupe Craze is $\frac{1}{6}$ of a pie. How many slices can Pete get from 15 of these pies?

5. Your music teacher treated everyone in the school choir to $\frac{5}{12}$ of a pie. It took 20 pies to feed all of you. How many students were in the school choir?

Critical Thinking Skills

Division Riddles

Use what you know to solve these division riddles.

1. What is my quotient? My dividend is 68,850. I like to be divided by a multiple of 10. My divisor is a product of 9. You will have the right quotient if each digit is 1 less than its neighbor on the left.

2. What is my divisor? My quotient is 17 with a remainder of 47. But my dividend is 1 less than 1000. Put me together, and you will see that my divisor is the product of 7 x 8.

3. What is my dividend? I am an even 2-digit number. The digit in my tens place is 7 higher than the digit in my ones place. If you add my divisor's two digits together, you will get 5. But 5 is still 1 more than the quotient, which is an even number, too.

4. Find the number of buttons in a sewing box. The number is more than 40 but less than 80. When the number is divided by 5, the remainder is 2. When the number is divided by 7, the remainder is 4.

Divisibility Moves On

Critical Thinking Skills

1. Which numbers are divisible by 3?

   51   63
         52
   67   313
         333
   818181   76543

A number is divisible by 3 if the sum of its digits is divisible by 3.

A number is divisible by 6 if it is divisible by 2 & 3.

2. Which numbers are divisible by 6?

   2570   96
   3429   880
   8372   412
   5144   683
   54   90

3. Circle the numbers that 5 divides evenly.

   11768   3745
   20040   630   27
   895   1989   382917
   7788992 19263054 1989

A number is divisible by 5 if it ends in 0 or 5.

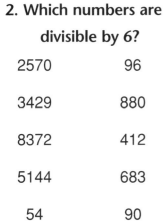

Critical Thinking Skills

The Divisibility Bowl

It's the Dodging Digits verses the Nuisance Numbers. But who's on which team, and who will win? Use the divisibility tests to find out.
Use page 73 to help you with the rules.

What you need to do:

1. Look at the players' numbers.
2. Any player wearing a shirt that has a number that is divisible by 3 is on the Dodging Digits team. Put a box around them.
3. Any player wearing a shirt with a number that is divisible by 4 is on the Nuisance Numbers team. Put an X on their shirts.
4. There are 2 referees wearing numbers that are divisible by 5. Put a circle around them.
5. Add up all the numbers on the referees' jerseys.
   If the sum is divisible by 3, the Dodging Digits win the game.
   If the sum is divisible by 4, the Nuisance Numbers win.

Who won the game? _____

Answer Pages

Page 3
1. quotient
2. dividend
3. dividend
4. divisor
5. quotient
6. dividend

Page 4
1. 2
   8
   7
   1
   5
   3
   6
   10

2. 4
   1
   12
   2
   3
   5
   7
   6

3. 2
   5
   4
   8
   1
   7
   3
   6

Page 5
4. 1
   3
   7
   9
   2
   11
   4
   5

5. 1
   2
   6
   4
   3
   9
   12
   11

6. 1
   4
   2
   9
   3
   12
   7
   5

Page 6
7. 1
   3
   4
   9
   10
   7
   2
   12

8. 1
   7
   4
   9
   6
   8
   12
   5

9. 1
   4
   3
   6
   2
   8
   5
   10

Page 7
10. 1
    7
    5
    2
    6
    4
    9
    3

11. 1
    5
    8
    9
    2
    6
    4
    3

12. 1
    4
    7
    5
    12
    8
    2
    3

Page 8
1. 4
2. 7
3. 6
4. 12

Answer Pages

Page 9
1. 7
   70
   700
2. 4
   40
   400
3. 4
   40
   400
4. 30
   300
   3,000

Page 10
1. 2
   20
   200
   2,000
2. 6
   60
   600
   6,000
3. 9
   90
   900
   9,000
4. 3
   30
   300
   3,000
5. 10
6. 70
7. 200
8. 40
9. 50
10. 70
11. 900
12. 500

Page 11
1. 2r1
2. 2r2
3. 7r2
4. 2r1
5. 3r1
6. 3r5
7. 8r2
8. 2r2
9. 8r3

Page 12
1. 7r1
2. 6r3
3. 7r1
4. 8r1
5. 5r2
6. 6r2
7. 8r3
8. 4r1
9. 1r1
10. 8r2

Page 13
1. 2
2. 4
3. 2
4. 4

Page 16
1. 13
2. 28
3. 14
4. 12
5. 29
6. 12
7. 12
8. 14

Page 18
1. 12r1
2. 12r4
3. 32r1
4. 11r3
5. 22r1
6. 12r3
7. 9r3
8. 22r1
9. 15r3
10. 13r1

Page 19
1. D
2. E
3. B
4. C
5. F
6. A

Page 20
1. 57
2. 328
3. 126
4. 35
5. 64
6. 108

Page 21
1. 195r2
2. 105r3
3. 174r2
4. 70r4
5. 246r1
6. 93r1
7. 416r1
8. 52r3

Page 22
1. 300r1
2. 208r3
3. 309r1
4. 40r4

Page 23
1. 1r8
2. 4r2
3. 3r9
4. 3r8
5. 3r4
6. 8r5

Page 24
1. 7r6
2. 8r7
3. 9r18
4. 4r54
5. 7r35
6. 8r10
7. 2r10
8. 5r89
9. 5r6
10. 2r9
11. 2r3
12. 4r14

Answer Pages

Page 25
1. 20
2. 20
3. 90
4. 20
5. 70

Page 26
1. 3
2. 4
3. 4
4. 7
5. 5
6. 6
7. 3
8. 5
9. 6
10. 7
11. 5
12. 5

Page 27
1. 3r14
2. 4r2
3. 2r5
4. 4r11
5. 5r6
6. 2r19
7. 4r2
8. 6r7
9. 4r11
10. 2r4
11. 2r5
12. 3r2

Page 28
1. D
2. I
3. A
4. J
5. C
6. H
7. E
8. B
9. G
10. F

Page 29
1. 20
2. 20
3. 230
4. 90
5. 350

Page 30
1. 22
2. 29
3. 36
4. 52
5. 64
6. 20
7. 19
8. 46
9. 25
10. 27
11. 12
12. 18

Page 31
1. 31r5
2. 25r19
3. 20r7
4. 19r49
5. 19r3
6. 27r2
7. 13r26
8. 58r4
9. 16r8
10. 10r79
11. 10r86
12. 37r8

Page 32
Across
1. 52
3. 31
5. 79
6. 49
7. 63
9. 28
11. 21
12. 34

Down
1. 57
2. 29
3. 34
4. 19
7. 62
8. 31
9. 23
10. 84

5	2	3	1
7	9	4	9
6	3	2	8
2	1	3	4

Page 33
1. 4 boxes. There will be 10 fliers in the last box.
2. 59 cartons. There will be 2 eggs in the last carton.
3. 19 crates
4. 47 pounds. No, there will be no wool left over.

Page 34
1. $0.50
2. $0.44
3. $0.43
4. $1.60
5. $0.92
6. $1.25

Page 35
1. $1.41
2. $0.69
3. $15.36
4. $0.93
5. $1.14
6. $0.40
7. $0.47
8. $0.40
9. $1.07
10. $0.50
11. $2.23
12. $2.68

Answer Pages

Page 36
1. more than $1.00
2. more than $1.00
3. less than $1.00
4. less than $1.00
5. less than $1.00
6. more than $1.00
7. less than $1.00
8. more than $1.00
9. more than $1.00
10. less than $1.00

Page 37
1. 309r10
2. 210r13
3. 205r11
4. 410
5. 32r36
6. 482

Page 38
1. 45
2. 35r15
3. 84
4. 88r26
5. 76
6. 48r8
7. 330
8. 72
9. 200
10. 234

Page 39
1. 74
2. 214r25
3. 194r43
4. 269r16
5. 178
6. 153r33
7. 396
8. 437
9. 89r18
10. 130r42

Page 40
1. 205r13
2. 406r54
3. 189r55
4. 2,674r14

Page 41
1. 765r42
2. 348r52
3. 2,466
4. 692r4
5. 552
6. 1,090r77
7. 3,269r6
8. 563r19
9. 951r46
10. 1,243

Page 42
1. not possible
2. possible
3. not possible
4. possible

Page 43
1. 167
2. 251r138
3. 87r154
4. 229

Page 44
1. 59r67
2. 286
3. 354r105
4. 67r42
5. 59r263
6. 73r162
7. 407r142
8. 94r761

Page 45
1. 860
2. 609
3. 866r682
4. 929r78

Page 46
1. 1,400
2. 908
3. 546
4. 865r495
5. 647r14
6. 705

Page 47
1. 16.2
2. 0.574
3. 4.381
4. 5.981
5. 3.55
6. 3.6625

Page 48
1. 45.32   4.532   0.4532
2. 0.7102   0.07102   0.007102
3. 98.603   9.8603   0.98603
4. 20.6   2.06   0.206
5. 54.59   5.459   0.5459
6. 160.583   16.0583   1.60583
7. 3.625   0.3625   0.03625
8. 68.24   6.824   0.6824
9. 0.2068   0.02068   0.002068
10. 1.3   0.13   0.013

Page 49
1. 3.8
2. 16
3. 7.52
4. 450
5. 7.3
6. 5.8

Page 50
1. 3.5
2. 17
3. 0.003
4. 20.5
5. 2.14
6. 2.5
7. 1.6
8. 90
9. 2.2
10. 0.02

Answer Pages

Page 51
1. 6.736
2. 4.07
3. 2.101
4. 1.929
5. 4.20
6. 5.24
7. 111.4
8. 0.6
9. 230.0
10. 5.3
11. E
12. B
13. C
14. D
15. A

Page 52
1. $2.65
2. 2.562 m
3. $25.14
4. 2.3 pounds

Page 53
1. 8
2. 1 3 3
3. $\frac{1}{4}$ 20
4. $\frac{8}{1}$ $\frac{7}{1}$ 56
5. $\frac{1}{4}$ $\frac{4}{3}$ $\frac{16}{3}$ $5\frac{1}{3}$
6. $\frac{12}{1}$ $\frac{5}{2}$ $\frac{60}{2}$ 30

Page 54
1. $1\frac{3}{4}$ 3
2. $14\frac{2}{3}$ $\frac{44}{3}$ 3
3. $7\frac{1}{2}$
4. 24
5. 18
6. $18\frac{2}{3}$
7. $31\frac{1}{2}$
8. $23\frac{1}{3}$
9. $7\frac{1}{2}$
10. $2\frac{6}{17}$

Page 55
1. 12 25 $2\frac{1}{12}$
2. $3\frac{5}{9}$
3. 9
4. $\frac{5}{8}$
5. 5
6. $2\frac{1}{4}$
7. $6\frac{2}{5}$
8. $1\frac{2}{13}$
9. $2\frac{13}{18}$
10. $5\frac{5}{8}$

Page 56
1. 18
2. $1\frac{7}{30}$
3. $5\frac{5}{6}$
4. $\frac{64}{71}$
5. $1\frac{1}{27}$
6. $3\frac{1}{2}$
7. $1\frac{3}{29}$
8. $3\frac{9}{13}$
9. $1\frac{1}{3}$
10. $1\frac{1}{9}$
11. $\frac{7}{15}$
12. 40

Page 57
1. 6
2. 2
3. 4
4. 3
5. $1\frac{7}{25}$
6. $\frac{8}{65}$
7. $\frac{1}{4}$
8. $\frac{3}{8}$
9. $\frac{1}{4}$
10. $\frac{1}{5}$

Page 58
1. 20%
2. 25%
3. 75%
4. 57%
5. 66%
6. 40%
7. 10%
8. 45%
9. 83%
10. 50%

Page 59
1. 4
2. 15
3. 5
4. 0.7
5. 6
6. 3

Page 60
1. I
2. F
3. D
4. C
5. H
6. E
7. B
8. G
9. J
10. A

Page 61
1. 3
2. 0.8
3. 1.1
4. 30
5. 2.5
6. 0.15
7. 3.6
8. 5.8
9. 7
10. $16.00

Page 62
1. 4
2. 2
3. 5
4. 7
5. 2
6. 4
7. 7
8. 4
9. 20
10. 4
11. 2
12. 4

Answer Pages

Page 63
1. 2 minutes, 12 seconds
2. 1 day, 2 hours
3. 3 hours, 9 minutes
4. 3 days, 6 hours
5. 1 year, 4 months
6. 9 months, 3 weeks
7. 7 weeks, 1 day
8. 5 minutes, 14 seconds
9. 2 days, 6 hours
10. 2 years, 5 months

Page 64
1. 3
2. 3
3. 3
4. 5.05
5. 18
6. 2

1. 4
2. 15
3. 120
4. 30

1. 2
2. 8
3. 12
4. 25

Page 65
1. 8
2. 4
3. 19
4. 3
5. 7

Page 66
1. D
2. B
3. F
4. E
5. A
6. C

Page 67
1. 9   846
2. 7   385
3. 6   282
4. 9   828
5. 7   441
6. 7   203

Page 68
1. 6
2. 16
3. 4
4. 15

Page 69

3-digit number	239	172	363	Answers will vary
Numbers with digits repeated	239,239	172,172	363,363	Answers will vary
÷ 13	18,403	13,244	27,951	Answers will vary
÷ 11	1,653	1,204	2,541	Answers will vary
÷ 7	239	172	363	Answers will vary

Page 70
1. 20¢
   16¢
   17¢
2. Frosty Bran Flakes
3. $7.90, 2 boxes

Page 71
1. 72
2. 42
3. 14
4. 90
5. 48

Page 72
1. 765
2. 56
3. 92
4. 67

Page 73
1. 51   63   818181
2. 96   54   90
3. 3745   20040   630   895

Page 74

Dodging Digits win!